SPEEDBIKES

SPEEDBIKES

Mick Woollett

B. T. Batsford Ltd. London

ISBN 0 7134 1294 1 (limp)

Typeset by Tek·Art Ltd London
and printed in Hong Kong
for the publishers
B. T. Batsford Ltd.
4 Fitzhardinge Street
London W1H 0AH

CONTENTS

Introduction

Annually, the world's motor cycle industry pours millions of pounds into the design, development and manufacture of competition motor cycles – and millions more paying top riders to get the best out of them.

The aim is to build race winners – machines that are superior to any others in their class. For winning races not only proves that a certain factory has managed to produce the best machine in a particular competition – it also generates publicity which the factory hopes will lead to increased sales of their bread-and-butter road machines.

In this book, *Motor Cycle Weekly* editor, Mick Woollett, who reported the Grand Prix and international Motor Cross meetings for over 20 years, takes a technical look at the current crop of top road racing, moto cross and drag racing machines.

Historical background and technical innovation are highlighted together with the facts and figures relating to each machine profiled – from the 50cc Kreidler road racer to the mighty 1123cc Honda Superbike racer and on to the championship-winning 500cc Suzuki moto crosser and the fastest dragsters in Europe. All are power bikes – built with the aim of winning.

LIGHTWEIGHTS AND LIGHTER BIKES

Armstrong CM36

The British Armstrong company came into motor cycle racing in 1981 with a 250cc machine powered by an Austrian Rotax twin-cylinder two-stroke engine. In 1982 they branched out into the 350cc class with a similar machine but this time powered by their own engine – designed by their director of engineering, Barry Hart.

Raced by Australia's Jeff Sayle and Clive Horton the CM36 proved fast enough to finish in the first ten in world championship races in its first year, a promising start.

Power unit is a water-cooled, two-stroke twin with the cylinders positioned one behind the other to reduce frontal area. Bore and stroke are 64 × 54mm to give an exact capacity of 347cc. Maximum revs are 10,800 and the power output is over 80 bhp – enough to give the Armstrong a top speed of around 150mph.

Lubrication is by four per cent oil in the petroil and under normal racing conditions the Armstrong covers just 15 miles per gallon.

Designed by Mike Eatough, the frame is conventional with a rocker arm rear suspension layout controlled by a single gas-oil Armstrong unit. Front fork is an Italian Marzocchi and 18in. wheels are fitted front and rear.

Total weight, ready to race but without fuel, is 238lb and the machines are finished in an eye-catching livery of yellow and blue with silver lining. Replica models can be bought from the Bolton factory for around £5000.

Bartol WMB 250 and 350

Working with the aid of two helpers, Austrian Harald Bartol, an accomplished Grand Prix competitor who retired from big time events in 1981, has produced two of the fastest machines in the 250 and 350cc classes – the disc valve, water-cooled, twin-cylinder two-strokes raced by the French ace Patrick Fernandez throughout the 1982 season.

Bartol first became fascinated by the two-stroke engine when he worked on his own machines seeking to increase their power by developing special cylinder barrels. The porting of these, coupled to carburation and exhaust systems, is all-important to two-stroke performance.

Not content with the engines available, Bartol set out, in 1977, to build his own; a time-and-money-consuming business with no guarantee of success – particularly when he was up against the might of the Japanese manufacturers.

Bartol stuck to an orthodox layout with the cylinders (54 × 54mm for the 250cc and 64 × 54mm for the 350cc) side-by-side. Carburettor sizes are 36mm for the smaller engine and the same basic Mikuni bored out to 37.5mm for the bigger unit.

Maximum revs are 12,000 and 11,000 and Bartol reckons the power to be in the region of 70bhp for the 250cc and close to 90 for the 350cc – giving top speeds of 145mph and 155mph.

Having carried out a lot of experiments he has cut the ratio of lubricating oil in the petrol to three per cent (33 to one) and uses a special brew of half

synthetic oil and half castor-based oil. Gearbox is a six-speeder.

Frames are by Dutchman Nicco Bakker fitted with Yamaha racing forks and rocker arm rear forks controlled by Swiss ATZ units. Wheel sizes are 16in. front and 18in. rear with Italian Brembo disc brakes and calipers.

With rider Fernandez sponsored by the French Pernod drink company, the bikes are painted Pernod yellow, white and blue.

Garelli 125

Spain's veteran lightweight star Angel Nieto won his eleventh world championship on a works 125cc machine fielded by the Italian Garelli factory which made an impressive come-back to racing in 1982 after years out of the sport.

In fact Garelli cut a few corners by signing on Dutch two-stroke expert Jan Thiel, the brain behind several race-winning machines in recent years including Bultaco and Minarelli. No matter, it was Garelli who provided the finance that allowed Thiel to continue to develop his machines.

With Nieto riding, backed by team-mate Eugenio Lazzarini, Garelli dominated the 125cc world championship Grands Prix that year finishing first and second despite a strong challenge by Sanvenero and MBA.

The Jan Thiel-designed engine is a parallel twin two-stroke with the cylinders inclined 35 degrees forward. Bore and stroke are 44 × 41mm to give an exact capacity of 124.6cc. Compression ratio is 14 to one and output is 46bhp at 14,600 rpm – not far short of the power produced by the TT winning 500cc works Nortons of the early fifties.

Carburettors are 29mm Italian Dell'Orto instru-

ments, and induction is controlled by disc valves. Gearbox is a six-speeder and the complete engine-gearbox unit, ready to race except for water in the cooling system, weighs just 50lb.

Garelli use a monocoque frame fabricated from alloy sheet; the main box section which doubles as the fuel tank and the main spine from which the engine-gearbox unit is suspended.

Front fork is a Marzocchi fitted with twin Brembo disc brakes with Garelli's own mechanical anti-dive mechanism. Rear suspension is controlled by a single Marzocchi unit and 18in. electron wheels fitted with Michelin tyres are used front and rear.

Ready to race, except for fuel, the Garelli weighs just 170lb and has a top speed of about 130mph.

Kawasaki 250 and 350

The 250cc and 350cc Kawasakis with their unusual twin-cylinder engines with the bores set one behind the other, tandem-style, have won more solo world championship Grands Prix races during the past five years than any other make – and out of the ten world championships Kawasaki riders have won eight titles.

Yet these machines took a long time to mature. Kawasaki exhibited a prototype as long ago as the Amsterdam Show of 1970 and there were one or two early and not very impressive outings in America.

They were re-launched in 1977 in the 250cc class only and were an immediate threat with Mick Grant winning the class at the Dutch TT and going on to

score a second success in Sweden. After a year of development the Kawasakis really came good in 1978. During the previous winter the factory had signed South African Kork Ballington to ride and he responded by winning both the 250cc and 350cc world championship in 1978 and again in 1979.

Serious stomach trouble put him out of contention the following year but West German Anton Mang won the quarter-litre class for Kawasaki and in 1981 the young Bavarian put them back on top in both categories with a clear-cut double.

Concentrating on their 500cc challenge, Kawasaki failed to do any real development work on the machines for the 1982 season but Mang still managed to hold onto the 350cc crown.

Externally the two engines are identical – apart from castings on the cylinder barrels which give the capacity. In fact the only difference is the size of the cylinder bores. The 250cc engine has a bore and stroke of 54×54mm to give an exact capacity of 247cc; the larger engine has a 64mm bore (347cc).

The two-stroke barrels have five transfer ports and induction through the Mikuni carburettors (34mm on the smaller engine, 36mm on the larger) is controlled by disc valves. Gearboxes are six-speeders and the rider can juggle with no less than four different ratios per gear as he strives to get the best gearing for particular circuits.

Maximum power from the 250cc is around 70bhp to give a top speed of over 140mph while big brother whacks out some 90 horse power and is capable of about 150mph.

Kawasaki were the first to perfect the now almost universal rocker arm rear suspension, where the movement of the rear wheel is controlled by a single suspension unit linked to the rear swinging fork by a system of rocker arms. These can be adjusted to change the rear springing – stiffening it or making it softer as required. A Kyaba gas unit is fitted. At the front Kawasaki forks take the load and braking is by a single disc front and rear.

Kreidler 50 Van Veen GP

Weighing only 110lb the 50cc Kreidler on which Switzerland's Stefan Dorflinger won the 1982 World Championship is capable of speeds of over 120mph in good conditions – but obviously strong winds and steep gradients affect the performance of these lightweight machines far more than they do the bigger bikes.

Built by the West German Kreidler company and developed over many years by their Dutch importer van Veen, the Kreidlers have always been the mainstay of 50cc racing, producing small batches of machines for sale from time to time.

The compact engine-gearbox unit is mounted with the water-cooled, two-stroke cylinder jutting forward horizontally. Bore and stroke are 40×39.7mm (exact capacity 49.9cc) and the power output is 20bhp at 16,000rpm. A disc valve controls induction from the 28mm Mikuni carburettor.

Cut into the tiny cylinder barrel are four transfer ports and one exhaust. The piston has a single plain ring and both big and small ends run in needle roller bearings. The water in the cooling system is circulated by a pump through the tiny radiator mounted ahead of the cylinder.

Gearing is vitally important on a 50cc racer – in fact until the number of gears was limited to a maximum of six the works 50cc racers had 12-, 14- and even 16-speed gearboxes! This allowed the tuners to increase peak power at the expense of a spread of power but it was clearly not in the interests of those developing roadster engines.

Now, to get the best out of the machine, Kreidler have three different sets of ratios for each of the six gears – the rider changing them after each practice session until he feels he has got the best possible combination of gears for that particular circuit.

Frame is basically a box spine made up of welded steel tubing. Front fork is Kreidler and both wheels are 18in. – the front shod with a 2in. section tyre, the rear with a 2.50 cover. A single hydraulic disc brake is fitted to each cast alloy wheel.

Morbidelli 125cc

Italian-built Morbidelli machines won the 125cc World Championship three times in succession in the seventies (1975 to 1977 inclusive) and for a while machines built by them for private owners were the best that you could buy for the class.

With backing from West German enthusiast, Michael Krauser, Switzerland's Stefan Dorflinger raced what is probably the fastest ever 125cc Morbidelli throughout the 1982 season, notching some notable successes competing against the works Garelli, Sanvenero and MBA (who took over the production of the original Morbidelli design) riders.

Designed by the famous German Jörg Möller, the engine is an orthodox side-by-side, near-vertical, two-stroke twin. It is water-cooled and has disc valves to control induction through the two 28mm Mikuni carburettors which are mounted on each side of the crankcase.

Bore and stroke are 44×44mm and maximum revs through the gears are 15,000 with a 14,500 limit in top (gearbox is a six-speeder). Horse power is about 40 – enough to give the Morbidelli a top speed of 135mph when timed on the long straight at the Paul Ricard circuit during the French Grand Prix.

Lubrication is by four per cent (25 to one) oil in the petrol and it is interesting to note that like most engines it lasts longest when using castor-based oil (oil which has a high percentage of natural oil from castor oil beans).

Fuel consumption is about 27 miles per gallon – surprisingly good for a racing engine. Frame is by Dutch specialist Nico Bakker and is of sturdy double loop construction, made from chrome steel tubing.

Front fork is an Italian Marzocchi while a single de Carbon unit controls the rear suspension – mounted almost horizontally in the style pioneered by Yamaha in their monoshock design. Both wheels are 18in. Campagnolo cast with Zanzani disc brakes (double on the front wheel) and Brembo calipers. Total weight, without fuel, is just 171lb – and if you wanted to buy a replica model the cost in 1982 was about £5500.

The main box section of the Garelli's frame houses the fuel tank, while the main spine contains the engine-gearbox unit

Waddon Ehrlich 250

After some fifteen years out of motor cycle racing Dr Joe Ehrlich returned to his first love when he joined forces with Waddon Engineering to produce the Waddon Ehrlich 250.

In the late fifties and early sixties the Austrian-born doctor of engineering had numbered some of the top names of racing among the riders of his trend-setting EMC (Ehrlich Motor Cycles) machines including Mike Hailwood and Phil Read. At one time his efforts were backed by Rolls-Royce and de Havilland.

He dropped out to concentrate on other ventures, but when Waddon appealed to him for help the temptation was too much and in December 1981 he announced his return.

Basis of the Waddon Ehrlich is an Austrian Rotax engine but this has been extensively modified in the Ehrlich works at Bletchley. Bore and stroke remain 54 × 54mm but special pistons, barrels, connecting rods and big-end bearings replace the originals.

Initial results were disappointing but then came a major triumph when Ireland's Con Law won the 250cc Junior TT in the Isle of Man, beating a host of Yamaha and Rotax powered rivals. It was 'Dr Joe's' first TT win. Hailwood had so nearly won the 125cc race for him back in 1959 only to break down on the last lap.

Law's TT win was quickly followed by new signing Graeme McGregor riding a works Waddon Ehrlich into second place in the 250cc class of the Belgian Grand Prix, beaten only by world champion Toni Mang (Kawasaki).

Revving to 12,500 and breathing through two 36mm Dell'Orto carburettors, the water-cooled in-line engine produces: 'well over 70bhp', according to Ehrlich. Lubrication is by four per cent castor-based oil in the petrol and gearbox is a six-speeder.

The Waddon frame is fitted with a Marzocchi front fork and has rising rate rear suspension controlled by a single de Carbon unit. Wheels are

Kawasaki riders Mang (left) and Balde (right) making final adjustments

15

18in. front and rear. Ready to race weight, less fuel, is 225lb and the colour scheme for the Ehrlich machines is silver.

Yamaha TZ250

In 1982 the Yamaha TZ250 celebrated ten years in production – and the little, water-cooled, twin-cylinder two-stroke remains the most popular production racing machine available for the class. Well prepared and ridden hard it it still a challenger for top Grand Prix honours but, equally important, it is the machine on which hundreds of riders around the world learn their racing and have their fun.

It was early in 1972 that Yamaha introduced the TZ250 – and the near identical TZ350. They replaced the air-cooled models which, since they first went on sale in 1969, had built up a tremendous reputation for speed and reliability.

The 1982 model (TZ250J – the J signifying the year of manufacture which started with A in 1972) is still powered by the relatively simple engine that has been the hallmark of the TZ models. The cylinders are set side-by-side. Bore and stroke are 56×50.7mm to give an exact capacity of 249.8cc.

Unlike the majority of rival machines, the Yamaha does not have disc valves. Instead it relies on piston-controlled ports – four transfer and one exhaust. Two 36mm Mikuni carburettors are fitted. The engine revs to 12,000 and produces around 60bhp to give a top speed of close to 140mph.

Gearbox is a six-speeder with multi-plate clutch. Frame is of simple double-loop layout with an aluminium box section rear swinging fork controlled by a single long suspension unit mounted under the aluminium fuel tank.

Front fork is Yamaha. Wheels are 18in. with twin Yamaha disc brakes at the front and a single Yamaha disc at the back. Weight, ready to race but without fuel, is 235lb.

Since 1972 Yamaha have produced over 2000 of the TZ racing models and have contributed enormously to the success of road racing around the world.

All set to go – Balde awaits the flag

To get the best out of the machine, Kreidler have three different sets of ratios for each of the six gears

19

The Morbidelli 125 engine, designed by Jörg Möller, has two 28mm Mikuni carburettors mounted on each side of the crankcase

Below   *The Waddon Ehrlich Team pose at Silverstone after their 1982 TT victory*

Opposite   *The Yamaha TZ250 – celebrating its tenth birthday and still the most popular machine available for its class*

FIVE HUNDRED PLUS

Ducati 600 Pantah

One of the few major motor cycle titles won by a European machine in 1982 was the Formula Two World Championship. This went to an Italian Ducati 600 Pantah, ridden by British veteran Tony Rutter, who won the class in the Isle of Man TT, and went on to clinch things at Vila Real in Portugal and at the Ulster Grand Prix.

Formula Two is based on road machines up to 600cc and although there are a number of high performance four-cylinder models in this bracket they simply could not match the all-round excellence of the Ducati which blends good usable power with light weight, good brakes and, above all, excellent handling.

The engine, designed by Fabio Taglioni, is a 90-degree vee-twin. It was first produced as a five-hundred and then, with an eye on Formula Two, as a six-hundred. Bore and stroke are well over-square (bore bigger than the stroke) at 80×58mm, to give an exact capacity of 583cc. With a compression ratio of ten to one, and breathing through 40mm Dell 'Orto carburettors, the engine produces over 60bhp when prepared for Formula Two racing.

Valve gear is desmodromic (the valves being closed as well as opened by cams) and the engine happily spins at 9000rpm and will go to 10,000 in the gears without damage. Gearbox is a five-speeder.

The engine-gearbox unit is mounted in a spine-type steel tube frame. The duplex lower rail run each side of the rear cylinder so that the unit is in fact suspended by the frame – a layout since adopted by Yamaha for their XZ550 sports vee-twin.

Both wheels are 18in. – the front carried in a Ducati telescopic fork and the rear in an old-style swinging fork controlled by twin suspension units. Twin 10in. discs are fitted to the front wheel with a single disc of the same size at the rear.

Even when stripped for racing, the Ducati is still quite a heavy machine, but the small frontal area of the vee-twin engine plus the good power gives it a top speed of 140mph.

Elf E

With financial help from the French Elf petrol company, André de Cortanze has spent most of his spare time over the last five years designing, building and developing the most revolutionary machine in motor cycle racing – the Elf E ('E' for experimental).

André, a top class enduro and trials rider before a serious injury put him out of the sport, works as an engineer with Renault – and his Elf racer incorporates many ideas from the car racing world.

Most obvious of these is the front suspension with hub centre steering – controlled by a single ATZ unit. Special emphasis has been placed on weight saving and the rear single-sided swinging arm is a magnesium casting, while the disc brakes are carbon fibre with Formula Two car racing AP calipers.

None of the high-performance 4-cylinder models can match the all-round excellence of the Ducati 600 Pantah

The most revolutionary machine in motor cycle racing, the Elf E incorporates many features from the world of car racing

Weight, with over a gallon of oil, is 396lb which André claims makes the Elf E the lightest endurance racer with all the rival machines over the 400lb mark.

To keep the weight low the fuel tank is under the engine. It is retained by four bolts and, with aircraft style quick release fuel pipes, the whole tank can be changed in three minutes. A Renault fuel pump is used to raise the petrol to the carburettors. With the fuel tank underneath, the exhaust pipes exit to the rear over the top of the engine.

There is no frame – front and rear suspension simply bolt onto the Honda RSC engine gearbox unit. In fact Honda were so impressed by the early prototype that they supplied Cortanze with a free 1062cc four-cylinder engine similar to the ones used by the factory riders.

The Elf E led the Bol d'Or 24-hour race briefly in 1981 but so far major honours have eluded the team, though the machine continues to attract the attention both of the public and the factories.

Honda Superbike

The largest capacity road racer ever built by Honda, the 1123cc Superbike version was produced specially for the British championship of the same name – and, brilliantly ridden by Ron Haslam, a Honda of this type took the title in 1981.

Engine is a conventional across-the-frame four-stroke, four-cylinder design, pioneered by the Italian Gilera and MV Agusta factories and first copied and

then improved by the Japanese. Bore and stroke are 72 × 69mm and maximum power of around 150bhp is developed at 9500rpm.

Normally peak revs in the gears is 10,000 but the engines can be revved far higher without damage – in fact riders claim to have seen rev counter readings of 14,000 when they missed a gear and revs shot sky high. This ability to over-rev without damage is mainly due to the four-valves-per-cylinder layout favoured by Honda. The relatively small and light valves are less prone to float and bounce than heavier and larger valves used in two-valve layouts like the rival Yoshimura Suzuki.

Lubrication is by just over a gallon of oil carried in a heavily finned sump. Four 33mm Keihin carburettors feed fuel to the engine and, covering about 20 miles per gallon, the big Honda is far more economical than the 500cc two-stroke Grand Prix machines.

The frame is the work of Ron Williams, Honda's UK-contracted designer who used to make the Maxton Yamaha racers. Rear suspension is rising rate controlled by a single Koni. Front fork is a Showa fitted with Ron's own design of mechanical anti-dive to prevent excessive front-end dipping under harsh braking from high speeds.

Wheel sizes are either 16 or 18in. front, depending on circuit and rider preference, and 18in. rear, both shod with Dunlop tyres. And despite weighing in at a hefty 390lb the Honda Superbike clocked 172mph through the speed trap at Ireland's North West 200 race in May 1982.

The Honda Superbike may weigh 390lb, but it is still capable of hitting 172 mph

Hondas FWS Vee-Four

The most impressive racing machine to make its debut in 1982 was the 1024cc Honda FWS Vee-four – the most powerful road racer ever built.

Developed for American road racing the FWS almost scored a win in its first race when Freddie Spencer rode it into second place in the Daytona 200, despite having to stop twice to change rear tyres because Michelin had supplied covers that were too soft and wore too quickly.

Team-mate Mike Baldwin went on to win the American road racing championship on a sister machine but plans for Joey Dunlop to race one in the Isle of Man TT in 1982 came to nothing after the gearbox failed in practice.

Engine of the FWS is a vee-four. Water-cooled, and with double overhead camshafts, the bore and stroke dimensions are way over-square at 78 × 53.6mm. Peak revs are an impressive 14,000 and power output is around the 150bhp mark – enough to propel FWS and rider at 170mph under favourable conditions.

The bank of four Keihin vacuum-type carburettors is mounted in the vee of the cylinders. The exhaust pipes from the front pair of cylinders exit under the engine into a silencer mounted low down on the right of the rear wheel. The two pipes from the rear cylinders exit from the rear of the block on each side of the single Showa rear suspension unit. They then join up under the seat and run to a silencer mounted in the right of the seat.

Two radiators are fitted. The main one to cool the water – the smaller one, mounted low down, to cool the near two gallons of oil in the lubrication system. Gearbox is a six-speeder and the clutch, worked by the normal handlebar lever, is hydraulic.

Right side of the steel tube frame can be unbolted to facilitate engine removal. The Showa front fork is fitted with a hydraulic anti-dive system and the twin disc Nissin front brake set-up has two pads each side of the calipers to give a total of eight pads acting on the front wheel.

The Honda FWS Vee-four can proudly claim to be the most powerful road racer ever built

27

Honda come up with another winner, the two-stroke NS500

Opposite At only 125bhp, it may not be quite as powerful as its rivals, but makes up for this deficiency by being smaller and lighter

Wheel sizes are 16in. front and 18in. rear and a safety feature pioneered by Honda is an over-run clutch. This means that if the engine locks up or if the rider changes down too quickly the rear wheel will not lock up as it normally would – instead the over-run clutch slips and the bike coasts along until engine speed and rear wheel speed match up.

Now a 750cc version of the FWS is being readied for European racing.

Honda NR500

Honda's four-stroke NR500 Grand Prix road racer is probably the most complicated and expensive machine ever developed for motor cycle sport – and must rank as one of the sports greatest failures. For, despite a massive development and racing budget, it failed to finish among the top ten in three seasons of world championship racing.

Rumour has it that when Honda decided to return to Grand Prix racing in the late seventies they were not aware that the Féderation Internationale Motocycliste (governing body of motor cycle sport) had restricted the number of cylinders permitted for racing to four.

This made it virtually impossible for a four-stroke engine to compete with the two-strokes – and explains why the Honda was so unsuccessful. For the machine they eventually wheeled out for the first time at the British Grand Prix at Silverstone in August 1980 was powered by an engine which appeared to be a vee-eight but was in fact a vee-four.

In an attempt to get round the rules, Honda built an engine with four cylinder bores and pistons but with eight connecting rods and, in effect, eight cylinder heads.

To do this they developed the most unorthodox cylinder barrels and pistons ever used. For, instead of being round, the bores are shaped like a running track with flat sides and rounded ends. To prevent the unusual shaped pistons wobbling about, two connecting rods are fitted.

The similar shaped combustion chamber has no less than eight valves. These are arranged in two groups of four with a sparking plug in the centre of each group. The water-cooled engine is mounted in the frame with one pair of cylinders jutting forward almost horizontally while the camshafts (two per pair of cylinders) is taken from the centre of the crankshaft. The eight Keihin flat-slide carburettors nestle in the vee between the cylinders, leaving the exhaust pipes to exit from the lower side of the front pair of cylinders and from the rear of the upright bank.

Sealing the strange shaped pistons is obviously a problem and Honda use three rings against the one of most two-strokes. Peak revs of the 1982 engine (the design of the unit has been changed several times) is a phenomenal 22,000rpm – but the NR500 still could not catch the two-strokes and so Honda, while keeping on with the development of this unusual machine, bought out their own two-stroke five-hundred, the NS500 and straight away got among the Grand Prix places, with Freddie Spencer taking third position in his first race on the new bike.

Honda NS500

After wasting three years and around £5m on trying to develop the four-stroke NR500 into a Grand Prix winner, and failing miserably – despite a succession of riders an NR500 never managed to finish in the first ten in a world championship race in three seasons – Honda did the sensible thing and produced a two-stroke for the 500cc class.

Labelled the NS500, the newcomer, despite a very unorthodox layout, proved an immediate success. In its first race, the Argentine Grand Prix of 1982, Freddie Spencer battled it out with Yamaha aces Kenny Roberts and Barry Sheene to finish a very close third – and later in the year he won the Belgian and San Marino rounds.

Unlike their rivals in the class, Honda did not opt for a four-cylinder engine – the maximum allowed under the present regulations. Instead they built a three-cylinder unit. Another surprise was that there were no disc valves – just piston-controlled ports with reed valves between the 36mm Keihin carburettors and the barrels to prevent blow-back.

Bore and stroke of the water-cooled cylinders (the outer two near vertical and the centre one jutting forward horizontally) is 62 × 55mm. Maximum revs are normally 11,000 with the riders taking the engine to 12,000 in the lower gears of the six-speed box.

Power at 125bhp is slightly down on Suzuki and Yamaha, but the Honda is smaller and, at 264lb, lighter than its rivals.

The welded aluminium frame is fitted with a single Showa rear suspension unit and a Showa front fork. Wheels sizes are 16in. front and 18in. rear with disc brakes. The Nissin double disc on the front wheel has twin pistons per caliper. Wheels are of Honda Comstar design and during the 1982 season carbon

fibre replaced the aluminium spokes to save weight – and Honda also experimented with carbon fibre rear swinging forks.

All three riders (American Freddie Spencer, Italian Marco Lucchinelli and Japanese Takazumi Katayama) preferred to use French Michelin tyres.

Kawasaki KR500

Seeking to capitalize on their successes in the 250cc and 350cc World Championship, Kawasaki decided to move up into the 500cc class in 1980 with the KR500 – the engine of which is a square-four two-stroke which could be said to be two of their 250s side-by-side.

Certainly Kawasaki were able to use many of the lessons learned, while developing their smaller machines, in the KR500. Bore and stroke are 'square' at 54 × 54mm (same as the 250cc) and maximum revs are 12,000 with real power from 8000.

The cylinder bores are plated by Kawasaki's own Electro-Fusion process and each has five transfer ports and one exhaust. Induction is controlled by disc valves and carburettors are 34mm Mikuni. Lubrication is by the usual crude but effective petroil system with oil simply mixed in with the fuel – at the ratio of 25 to one (four per cent oil).

Maximum power is about 125bhp and this is transmitted via a six-speed gearbox to the 18in. rear wheel. Front wheel is 16in. mounted in a conventional Kyaba telescopic front fork with coil

Full of surprises – Honda built a three- instead of a four-cylinder unit and there are no disc valves

31

The Kawasaki KR500 is the only machine of its class to use a mechanical anti-dive system to prevent the front suspension dipping under heavy breaking. Kork Ballington riding

springs and oil damping. Alone among the factory five-hundreds, Kawasaki use a mechanical anti-dive system which prevents the front suspension from bottoming under heavy braking.

At the rear a triangulated aluminium box section swinging fork is connected by rocker arms – to give rising rate suspension – to a single, vertically mounted Kyaba gas unit. Kawasaki disc brakes, twin on the front wheel, single at the rear , with magnesium calipers take care of the stopping and the Dymag cast wheels are shod with Dunlop tyres.

The frame is unusual. Originally it was a true monocoque of welded aluminium with the fuel tank an integral and stressed member. However, for 1982, this was replaced and while the basic layout was retained the tank is now separate.

Unfortunately, the KR500 has not been a great success. Progress was hampered by injuries to riders Kork Ballington of South Africa and Gregg Hansford of Australia. Hansford in fact retired at the end of the 1981 season and was not replaced. This meant that Ballington had to soldier on on his own and although he achieved some good results – and looked set to win the British Grand Prix at Silverstone in 1981 until a disc valve broke – he could not match the might of the other Japanese factories who fielded two and three works machines.

Kawasaki Endurance

Kawasaki are very much the kings of the long-distance endurance racing world, dominating the ultra tough events which make up the World Championship series, including three 24-hour events – the French Bol d'Or, the Montjuich event held in a park in Barcelona, Spain and the Liège race which is run over the Spa-Francorchamps circuit in Belgium.

The engine is basically a sports roadster, air-cooled, four-cylinder, double overhead camshaft unit enlarged by the factory to 1150cc (bore and stroke 74

Kawasaki know how to put together a machine strong enough to face the toughest events the World Championship series can offer – the Kawasaki Endurance

× 64mm). This develops a great deal of power low down the rev range to give the Kawasaki tremendous acceleration plus a top speed of around 160mph.

Gearbox is a five-speeder and maximum revs are 9000 – quite low by modern standards but when an engine has to run without breaking for 24-hours a low revving unit is obviously a big advantage.

Over a gallon of oil lubricates and helps to cool the engine and, to keep the temperature of the lubricant down, an oil radiator is fitted in the nose of the

streamlining above the twin headlamps. Fuel consumption works out at about 15 miles per gallon and during a 24-hours race one of these machines burns some 160 gallons of petrol – costing £280 at 1982 prices!

Frame is an orthodox duplex tube design with Kawasaki front fork fitted with the mechanical anti-dive system first used on the factory KR500 Grand Prix racer. Rear suspension has changed in three stages over the years. First came a normal twin shock

layout. This was replaced in 1980 by an unusual set-up where the movement of the strongly-braced rear fork was controlled by a single massive gas shock mounted on the off-side of the fork. Then in 1982 Kawasaki got up to date with a rising rate set-up with a single, centrally mounted shock.

Machines entered by the factory's French importer won the endurance championship in 1981, and again in 1982, despite major competition from Honda.

Suzuki RG500 Gamma

The Suzuki RG500 Gamma that carried Italian Franco Uncini to world championship honours in 1982 was the latest in a line that started back in 1974 when the Japanese factory first wheeled out a four-cylinder 500cc racing machine.

In those days a youngster named Barry Sheene was their Number One rider and Barry went on to win the world championship for Suzuki in 1976 and again in

Top 500cc Grand Prix machine in 1982 – the Suzuki RG500 Gamma

35

You'll certainly know when a Suzuki Formula One is around – capable of 11,000 revs, it's not easy to ignore

1977. Yamaha took the title, thanks largely to the brilliant riding of Kenny Roberts for the next three seasons in succession.

Then Marco Lucchinelli and Suzuki wrested the 500cc title from them in 1981 and when Lucchinelli was tempted away by Honda money it was Uncini who took over as leader of the Italian-based Gallina team to keep Suzuki at the top of the incredibly competitive 500cc racing scene.

Basically the machine is surprisingly similar to the original 1974 model. The engine is a two-stroke, with the cylinders arranged in a square. Bore and stroke are 56×50.6mm and revving to 12,500 the power output is around the 130bhp – enough to propel bike and rider at speeds up to 160mph.

Induction is controlled by disc valves. Carburettors are four 36mm Mikunis and fuel consumption is around 12 miles per gallon! Lubrication is simply by oil being mixed with the petrol at a ratio of one part of oil to 25 of petrol.

The gearbox is a six-speeder and an interesting point about the design is that the gearchange shaft projects from both sides of the gearbox so that the gear pedal can be fitted to either side to suit the rider's preference.

While the engine is similar to the original, the frame is entirely new. Instead of the tubular steel frame of previous years, Uncini used a square section welded alloy frame with Suzuki anti-dive telescopic front fork and rising rate rear suspension controlled by a single gas-air unit.

Cast alloy wheels (16in. front, 18in. rear) shod with

Roger Marshall, Suzuki's top rider in Formula One, in action

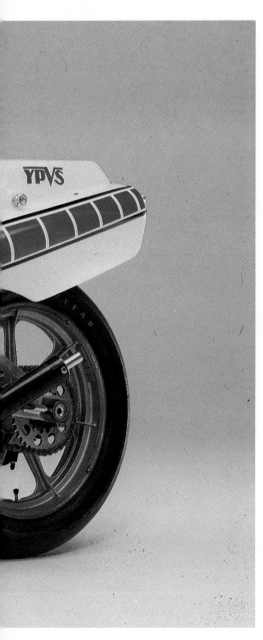

Michelin tyres are fitted with twin disc brakes at the front and a single disc at the rear. Total weight of the machine ready to race, except for fuel, is 275lb.

Suzuki Formula One

One of the finest sounds in motor cycle racing is that of a 1000cc four-cylinder, four-stroke Suzuki Formula One engine on full song. For, at 11,000 revs, the noise leaving the single collector box into which the four individual exhaust pipes run is sheer music to the enthusiast's ear.

Developed for the Suzuki factory in Japan by the legendary Japanese tuner Pops Yoshimura, the big four-stroke Suzuki engine, based on the roadster GS1000 unit, is available in two sizes – a 997cc engine for European Formula One racing which is limited to 1000cc (750cc from 1 January 1984) and a 1023cc unit for American events where the limit is 1025cc.

Fitted with the latest type of twin plug per cylinder head, the smaller engine gives 138bhp while the larger produces an additional 10 bhp. Despite extensive experiments with four-valve heads, Yoshimura sticks to the simpler two-valve layout and the compression ratio at 9.1 to one is surprisingly low.

Yoshimura and his men have concentrated on developing a really flexible engine and the big Suzukis pull from well down the rev range with the power starting at 3500rpm. This makes the machines easy to ride and gives them tremendous acceleration despite their weight of around 360lb – some 80lb heavier than a Grand Prix five-hundred.

Carburettors are changed to suit the type of circuit. On short courses, where acceleration is more important than top speed, 31mm Mikunis are fitted. On longer circuits, where flat out speed is wanted, they are replaced by 33.4mm Mikunis.

For 1982 a monoshock rear suspension frame was introduced. But unlike the Grand Prix bikes the Formula One machines retain an orthodox steel tube frame. A Suzuki telescopic front fork is fitted and

Unable to compete with the power of their rivals, Yamaha produced their own square-four design – first the OW54

both wheels are 18in. with Suzuki disc brakes and cast alloy wheels.

In 1982 Suzuki's top rider in Formula One was Roger Marshall who won the big class at the Ulster Grand Prix and set a new lap record for the famous Dundrod circuit in Northern Ireland.

Yamaha YZR500 Square-Four

Barry Sheene's mounts in the 500cc Grands Prix in 1981 and 1982 were the factory Yamaha YZR machines – first the OW54 and then the almost identical OW60 introduced for 1982.

Sheene put up a string of impressive performances on these machines, which are powered by water-cooled, disc valve, two-stroke engines with the cylinders set in a square. These engines follow the design pioneered by the rival Suzuki factory who first introduced the layout to the 500cc class in 1974 (when ironically Sheene was riding for them).

Yamaha Number One and three-times World Champion, Kenny Roberts, used an OW54 throughout the 1981 season but Sheene had to wait until the fifth round, the French Grand Prix, until he got his. After that Sheene actually outscored Roberts in the remaining Grands Prix and ended the year with a win in Sweden.

He was having an excellent year on the replacement OW60 in 1982 until he switched to the vee-four OW61 for the British Grand Prix at Silverstone and suffered an 160mph crash in practice that put him out of racing for the rest of the year.

From 1973 when they first moved up to the 500cc class, right through to the start of 1981, Yamaha had steadily improved their relatively simple in-line, across-the-frame, four-cylinder, two-stroke design. And riding these machines Roberts took the title from Sheene (then leading the Suzuki team) in 1978 and held it in 1979 and again in 1980.

Eventually the orthodox piston port, in-line engine simply was not powerful enough to combat the

Suzukis and with Honda and Kawasaki also challenging, Yamaha produced their own square-four – a design which simplifies the use of disc valves.

These rotate between the carburettor and the crankcase, sealing the latter when the piston is falling and preventing blow-back through the carburettor. Bore and stroke are 56 × 50.6mm. The engine revs to 12,000 and produces around 130bhp.

Mikuni carburettors are fitted (36mm) and the engine runs on the usual four per cent mix of oil in the petrol – using fuel at the rate of about one gallon per 12 miles covered. Gearbox is a six-speeder with alternative clusters available for different circuits. Top speed is in the 160mph region.

Following the trend, the frame is fabricated from square section aluminium alloy. Yamaha front fork with air and spring suspension and oil damping are fitted, while Yamaha stuck to their monoshock rear suspension layout with a single long gas unit mounted almost horizontally under the fuel tank. Cast Morris wheels (16 or 18in. front, 18in. rear) with Yamaha disc brakes complete the set-up.

Yamaha OW61

In an effort to combat the successful Suzuki RG500 Gamma, Yamaha first produced the square-four OW54 then the similar OW60 and, for the 1982 season, the vee-four OW61 factory racer.

After months of rumours Kenny Roberts raced the machine for the first time in the Austrian Grand Prix at the super-fast Salzburgring circuit. But although extremely fast, handling problems plagued the machine all year and Roberts was unable to match the consistency of Suzuki's Franco Uncini who went on to win the 500cc World Championship.

In fact probably the most famous incident involving an OW61 was Barry Sheene's horrific crash while practising for the British Grand Prix at Silverstone. Until then Roberts had been the only rider with an OW61. For Silverstone Yamaha gave

Sheene a similar machine and it was while he was setting the bike up for the race that Sheene crashed at 160mph when he struck a machine lying on the track.

The vee-four, water-cooled, two-stroke engine of the OW61 is mounted with one pair of cylinders jutting forward horizontally. The other pair project up towards the steering head. The 36mm Mikuni carburettors nestle in the vee of the engine which is rubber mounted in an unusual welded aluminium frame. Unique to Yamaha, a power-valve system is fitted. Electronically controlled, these valves reduce the size of the exhaust ports at low revs and this improves acceleration by broadening the spread of power.

Bore and stroke are 56 × 50.7mm and, revving to 12,000, the engine produces 135bhp to give a top

. . . and then its replacement, the almost identical, OW60

41

The Yamaha OW61, a vee-four, was the bike on which Barry Sheene had his 160 mph-accident while practising for the British Grand Prix at Silverstone. Sheene survived, so did the bike, but it is not without its problems

speed of around the 160mph mark on level ground.

Rear suspension is unusual. For, instead of a vertically or horizontally mounted unit, the rear shock absorber is positioned across the frame below the saddle. The rear swinging fork is connected to the suspension unit by a system of bell cranks.

Conventional telescopic forks using air and springs with oil damping are fitted up front and incorporate Yamaha's anti-dive system which is designed to stop the nose of the machine from dipping when the

massive twin disc front brakes are used.

American Morris cast alloy wheels are used, the rear equipped with a single disc brake. Wheel sizes are 16in. or 18in. at the front (depending on the riders preference) and 18in. at the rear, with Roberts using Dunlop tyres while Sheene preferred Michelin.

Ready to race, except for fuel, the OW61 weighs 286lb.

MOTO CROSS

Honda RC500M

Between them, Honda's European factory-supported moto cross riders, André Malherbe of Belgium and Graham Noyce of England, have won the 500cc World Championship three times in succession 1979 to 1981 inclusive – and for the 1982 championship both were equipped with works RC500M machines.

Power unit is an air-cooled, single-cylinder two-stroke. Honda have kept the bore and stroke a secret but we do know the exact capacity is 498cc and the factory have revealed that it develops maximum torque (pulling power) at 6000rpm and revs only to 6500.

Peak power is certain to be around the 60bhp mark but more important to the riders than out-and-out power is having a flexible engine that will pull from low revs. This makes the bike easy to ride, gives it good acceleration and cuts down on the number of gear changes a rider has to make during a race.

The Honda engineers have obviously done this, for the number of gears is four – one less than on some earlier models. An unusual feature of the gearbox, and one that seems certain to be copied by other manufacturers, is an over-run clutch.

This was originally developed for the road racing NR500 and the idea is that if for any reason the rear wheel wants to turn faster than the engine, it can – without spinning the engine faster. This is done by a clutch mechanism that can be pre-loaded so that the rider can get a certain amount of rear wheel braking by closing the throttle, but when it becomes too much and threatens to lock the wheel the clutch slips and the wheel keeps turning.

In moto cross this acts as a shock absorber, smoothing out the sudden loadings normally transmitted to the engine when the rear wheel hits the ground after a jump. It is also a safety feature if the engine seizes or a rider changes down too rapidly.

Frame is orthodox welded tube with a Showa front fork and Honda's Pro-Link rising rate set-up at the rear controlled by a Showa or Whitepower unit. Both wheels (21in. front/18in. rear) have 12in. of travel. Sad to say, Honda did not make it four in a row. Both riders missed out with injuries and Brad Lackey (Suzuki) broke the Honda grip in 1982.

Husqvarna 500CR

Sweden's Husqvarna factory – famous for their sewing machines and lawn-mowers as well as for motor cycles – was the last European concern to win the 500cc Moto Cross World Championship. That was back in 1974 and the rider was Heikki Mikkola.

Since then, the Japanese have taken over, but Husqvarna keep battling on and still produce some of the finest machines that private riders can buy. The 1982 500CR, virtually alone in the big class, still has a rear suspension controlled by twin-dampers. But these Swedish Ohlin dampers have been moved right

Honda have succeeded in creating a flexible engine and in minimizing the amount of gear changes a rider has to make during a race – the secret of moto cross

forward and are steeply angled to give a massive 13in. of movement to the 18in. rear wheel.

The front fork is telescopic, with just over 10in. of movement and the wheel is 21in. Frame is orthodox welded steel tubing and Husqvarna stick to a steel tube rear fork.

The single-cylinder, air-cooled, two-stroke engine has a bore and stroke of 86 × 84mm (exact capacity 488cc) which with a 9.5 to one compression ratio gives some 57bhp at 7500rpm. Carburettor is a huge 40mm Mikuni.

Lubrication is by the usual two-stroke method of oil in the petrol (three per cent /30 to one). After fitting earlier models with five- and even six-speed gearboxes, Husqvarna reverted to a four-speed transmission for the 500CR, tuning the engine for a spread of power – which makes more ratios unnecessary, rather than going for peak output.

Factory rider in the 500cc class in 1982 was Frenchman Patrick Fura who was sponsored by the Pernod drink manufacturers. But it proved a tough year for Fura and Husqvarna and they made little impact on the scene.

KTM 495MC

Beautifully engineered and strikingly finished in red, white and blue the Austrian KTM 495MC is one of the most impressive bikes in modern moto cross – but like the other European factories KTM have found competition offered by the Japanese fierce, and works rider Kees van der Ven had a tough time in 1982.

In the seventies KTM-mounted Russian, Gennady Moisseev, won the 250cc World Championship three times (1974, 1977 and 1978) and van der Ven finished third in the same division in 1981 before moving up into the big class on the factory 495MC.

Bore and stroke of the single-cylinder, air-cooled, two-stroke engine are 92.25 × 74mm to give a swept volume of 495cc. With a compression ratio of 11.8 to one the engine develops 56bhp at 6400rpm with maximum torque at 5700rpm.

Graham Noyce riding the Honda RC500M – injury prevented him from trying to make it four-in-a-row for Honda in the World Championship in 1982

Husqvarna were the last European factory to win the 500cc Moto Cross World Championship – in 1974. Despite Japanese domination of the class, Husqvarna still continue to make fine bikes

Carburettor is a German Bing (40mm) and using Elf special racing oil KTM have cut the percentage of oil in the petrol to as low as two (50 to one). Ignition is by a Spanish Motoplat electronic unit. Primary drive is by gears to a five-speed gearbox.

Frame is an orthodox welded steel, double-tube layout. It is fitted with an Italian Marzocchi fork which gives 12in. of suspension movement at the front and with KTM's own single shock system at the rear (dubbed Pro-Lever) which is controlled by a Dutch Whitepower unit and which also has 12in. of movement.

Front wheel is 21in. shod with a 3.00 tyre while the rear is 18in. with a 5.00 – normally Perelli covers are used. Drum brakes are fitted front and rear to the production machines but van der Ven experimented with a disc on the front wheel throughout 1982.

Total weight ready for the fray, except for petrol in the two-gallon tank, is 230lb – which made the KTM the heaviest of the 1982 Grand Prix 500cc machines.

Kawasaki KX500

Although the smallest of the big-four Japanese manufacturers, Kawasaki moto cross machines have built up an enviable reputation through the years for speed, reliability and handling, and the factory have pioneered several technical innovations.

Probably the most important of these is the modern style of rear suspension where a single, vertically mounted unit, postitioned just in front of the rear wheel, and connected to the rear fork by a system of rods, gives greatly increased wheel movement coupled to rising rate suspension (where the suspension gets progressively harder as the wheel moves up).

Called 'Uni-Trak' by Kawasaki this has been copied by most other manufacturers in recent years but the Kawasakis remain fine handling machines and this helped works rider Dave Thorpe to achieve many successes during 1982 on the development KX500 machine – including a fine second place in the British Moto Cross Grand Prix. In fact he proved

KTM also find themselves outclassed by the Japanese, but what splendid bikes they still produce. Kees van der Ven lives in hope . . .

David Thorpe's Kawasaki KX500 – he came second in the British Moto Cross Grand Prix in 1982

such a threat that he was poached by Honda for 1983!

Engine is a single-cylinder, air-cooled, two-stroke unit. Bore and stroke are 86 × 86mm to give a capacity of 499cc and to keep weight to a minimum, the aluminium cylinder is plated by Kawasaki's own Electro Fusion system which does away with the need for a steel liner.

The engine revs to 8000 and whacks out a lusty 57bhp. Lubrication is by four per cent oil in the petrol and, depending on gearing, the KX500 covers about 15 miles per gallon. Carburettor is a massive 38mm

Mikuni and there is a reed valve between carburettor and cylinder to prevent blow-back.

Gearbox is a five-speeder driving the 18in. rear wheel via a long and exposed chain. Ignition is Kawasaki's own pointless CDI system.

Suspension movement is a massive 12in. front and rear – double that of just a few years ago. Front fork is Kawasaki with a 21in. wheel and, unusual for moto cross, a hydraulic disc brake. Weight has been kept down to 226lb.

Maico MC490

Although a tiny factory in comparison with the Japanese giants, the West German Maico concern has a world-wide reputation in moto cross circles and their production machines still sell well despite tremendous competition.

Lacking the incredible amounts of money that the orientals have poured into the sport, Maico have not been able to afford the top riders in recent years and an outright World Championship win has always eluded them.

However, they were second in the 500cc title chase in 1971 when Sweden's Ake Jonsson was the rider and again in 1973 when German Fritz Bauer was the pilot. In more recent times they have impressed in the 250cc class with England's Neil Hudson runner-up in 1979 and Dutchman Kees van der Ven also second in 1980.

So the latest Maico, the MC490, has a pedigree and it is worth noting that, in addition to Hudson, England's other star moto crosser, Graham Noyce, also learnt the trade on a Maico before switching to Honda and winning the 500cc world title in 1979.

The latest machine, raced in the 1982 World Championship events by Italian Maurizio Dolce, is a full-sized five-hundred with bore and stroke of 86.5 × 83mm to give an exact capacity of 487cc (the previous big Maico was 77 × 83mm – 386cc).

Compression ratio is 12 to one and power output is 53bhp at 7000rpm. Carburettor is a 40mm Bing and lubrication is by four per cent oil in the petrol. Gearbox is a five-speeder and primary drive is by duplex chain.

Suspension movement is 12in. with a telescopic fork at the front and Maico's own version of the now almost universal single shock, rising rate suspension at the back. Wheels are 21in. front and 18in. rear both fitted with conical hub brakes. The fuel tank holds just over two gallons and the MC490 weighs in at 226lb.

Suzuki RN82

After challenging for 500cc Moto Cross World Championship honours on Honda and Kawasaki machines, American Brad Lackey eventually achieved his aim and took the title on a factory Suzuki in 1982.

In fact 1982 proved doubly successful for Suzuki with Lackey's team-mate André Vromans of Belgium taking second place in the table – the pair of them well ahead of third-place Neil Hudson (Yamaha).

The factory RM498, finished in the distinct yellow and blue Suzuki colours, is powered by a conventional air-cooled, single-cylinder, two-stroke engine with over-square dimensions of 89 × 80mm to give an exact capacity of 498cc.

Power output is close to 60bhp with the engine revving to 8500. Using synthetic oil, the ratio in the petrol has been cut from the usual four per cent (25 to one) to little more than two per cent – the actual mix being 40 to one. This is important because the less oil added to the petrol the higher the octane rating remains. This in turn allows a higher compression ratio to be used with consequent higher power output.

Carburettor is normally a 38mm Mikuni and as usual with this type of engine there is a reed valve between carburettor and cylinder barrel. The spread of power is so good that Suzuki have actually cut the number of ratios in the gearbox from five to four speeds. This makes the machines easier to ride because, the less gear changes a rider has to make in the course of a gruelling moto cross race, the more energy he has simply to control the machine.

Frame is of orthodox welded tube construction fitted with Suzuki's patent Full Floater rear suspension with progressive, rising rate linkage controlled by a single vertically mounted Kyaba unit. Front fork is also Kyaba and both wheels (18in. rear and 21in. front) have 12in. of suspension travel.

Simple hub brakes are fitted front and rear and it is

worth noting that the fancy cast wheels so popular in road racing have found no favours in moto cross. The spoked wheel combines strength with light weight and continues to be the best bet for moto cross where wheels are subject to more stress and strain than in any other form of motor cycle sport.

Yamaha YZ490K

After winning the 250cc Moto Cross World Championship for Yamaha in 1981 England's Neil Hudson moved up to the 500cc class. There, riding a factory prototype, he finished a fine third in the big

Another victim to the Japanese supremacy, the Maico factory is nonetheless very well thought of in the moto cross world

division – and Yamaha immediately combined the lessons learned into their production moto cross machine for 1983, the YZ490K.

The single-cylinder, air-cooled, two-stroke engine has a bore and stroke of 87 × 82mm (487cc) and develops 56bhp at 7000rpm. Lubrication is by 40 to one oil (two-and-a-half per cent) in the petrol. Carburettor is a 38mm Mikuni and gearbox is a good old-fashioned four-speeder.

New for 1983 is the rear suspension. Gone is the long unit mounted high up under the tank and extending from the steering head to the triangulated rear fork. This is replaced by a shorter unit mounted just above the rear swinging fork pivot and connected to the fork by an adjustable linkage to give true rising rate suspension. Not only is suspension improved but handling is better because the weight of the unit is now much lower.

Travel of the rear suspension is now a massive 13in. Wheel is 18in. fitted with a 5.20 Bridgestone tyre. Front fork is Yamaha's own with air-assisted springs with oil damping. Wheel is 21in. with 3.00 tyre and suspension travel is 12in.

Fuel tank capacity is just over two gallons and the big Yamaha moto crosser is finished in white with a red seat.

This is the bike that took second place in the Moto Cross World Championship table, ridden by André Vromans – second to team-mate Brad Lackey

Neil Hudson tried his hand at 500cc moto cross, and finished third in 1982. Seen here is his modified bike for 1983

DRAGSTERS

Yamaha Dragster

Fuel is normally 90 per cent Nitro and ten per cent methane fed into the inlets by a fuel injection system designed by Reisten. The engine normally burns two gallons per run, including the burn out when the rider heats the rear tyre ready for the run by spinning the wheel on the tarmac using the power of the engine. This works out to a fuel consumption of eight gallons for every mile covered – each run costs Reisten close to £20 for fuel alone! Peak revs are 11,000 and maximum power is about 300bhp.
stronger home-built ones, and the normal five-speed gearbox with an epicyclic two-speeder.

Hydraulically operated, the change is triggered by the rider thumbing a button about a third of the way down the strip. The power of the already heavily tuned engine is further boosted by the use of an American Magnusson supercharger which can deliver up to 30lb per square inch of boost.

Fuel is normally 90 per cent Nitro and ten per cent methane fed into the inlets by a fuel injection system designed by Reisten. The engine normally burns two gallons per run, including the burn out when the rider heats the rear tyre ready for the run by spinning the wheel on the tarmac using the power of the engine. This works out to a fuel consumption of eight gallons for every mile covered – each run costs Reisten close to £20 for fuel alone! Peak revs are 11,000 and maximum power is about the 300bhp.

To translate this massive power into acceleration and speed the simple frame is fitted with a massive 16in.-wide racing car style slick rear tyre. There is no rear suspension. Up front a normal roadster telescopic fork with double-disc brake and cast alloy wheel take the strain. The total weight of the bike, ready to go, is 550lb.

Kawasaki Dragster

Europe's most successful drag racer, with a career at the top which spans over ten years, Henk Vink has built a succession of Kawasaki-powered drag bikes on which he estimates through the years he has spent something like £250,000! Luckily he is a relatively wealthy man with a flourishing motor cycle business in his native Holland which provides a sound financial base for his hobby.

In the sixties Vink was a successful trials rider who won the Dutch championship several times. Then drag racing caught his attention and he switched from the slow world of the observed trial, where bikes seldom move faster than walking pace, to the slingshot drag scene which now produces the fastest machines used in motor cycle sport.

For several years Vink relied on twin-engined machines but in 1982 he followed the trend for lighter bikes and campaigned a newcomer powered by a single, supercharged, four-cylinder double overhead camshaft Kawasaki unit.

Over the years these engines, produced to power the Kawasaki sports roadsters, have gained a reputation for their rugged construction and even when power output is trebled they remain reasonably reliable – though as power goes up so many of the original parts are replaced by special components.

The Vink engine is bored and stroked to increase the capacity from the 1089cc of the standard unit to 1356cc. The two valves per cylinder set-up is retained but valve sizes are increased and special pistons, con-rods and crankshaft are fitted.

Camshafts too are very non-standard and to further boost the power a supercharger is fitted to force the rich mixture of nitro and methanol into the cylinders. Maximum revs are 11,000 and the engine produces over 300bhp during the eight-second run down the quarter-mile strip.

The standard gearbox is replaced by the usual drag racing set-up – an epicyclic two-speed transmission which Vink activates by simply pressing a button – no disengaging the clutch or closing the throttle.

Europe's fastest accelerating motor cycle. Stephen Reisten in action on the Yamaha Dragster, capable of doing the standing start quarter-mile in 7.72 seconds

The frame is a simple welded tube, one with no rear suspension. Rear wheel is a 12in.-wide slick. Front fork is Kawasaki with twin disc brakes. Total weight ready almost 500lb – and Vink can cover the quarter mile from a standing start in just eight seconds, crossing the finish line at over 180mph!

Henk Vink, in his tenth year at the top of drag racing, in action on his Kawasaki Dragster

INDEX